The 1998 National Security Strategy repeatedly cites global environmental issues as key to the long-term security of the United States. Similarly, U.S. environmental issues also have important global implications.[1] This paper analyzes current U.S. Policy as it pertains to global warming and climate change. It discusses related economic factors and environmental concerns. It assesses current White House policy as it relates to the U.S. military. It reviews the Department of Defense strategy for energy conservation and reduction of greenhouse gases. Finally, it offers recommendations and options for military involvement to reduce global warming.

BACKGROUND

Global warming is not a myth or voodoo ecology. The 1900's have been the warmest century in the past 600 years. The temperature continues to rise every year.[2] If this trend is not arrested or reversed, it will have a devastating impact on the quality of life and the economy. The President's policy is to set up national and international guidelines and goals with specific target dates, in order to slow or halt global warming and its effects. The overall goal is a higher quality of life and an economically favorable outcome.

Global warming can be defined as the gradual increase in temperature caused by the increased production of carbon dioxide (CO_2) that has occurred during the post-industrialized age. The increased CO_2 emissions are a result of an increasing population that has been burning vast amounts of coal and fuel for most of this century.[3] Also, the production and use of specific chemicals by industry is slowly eroding the ozone layer, the atmosphere's protective layer between the Earth and the sun. When the ozone layer is depleted, the Earth receives more ultraviolet rays from the sun. This causes an increased incidence of skin cancer, cataracts, and immune suppression, as well as a higher mean temperature on earth.[4,5]

In nature, the presence of CO_2 in the atmosphere keeps the temperature an average 60 degrees Fahrenheit. With steady increases in CO_2 emissions it is expected that the temperature will increase 2-6.5 degrees Fahrenheit by the year 2100.[6] If the temperature continues to increase, the sea level could rise 3 to 16 feet due to melting of portions of Greenland and the Antarctic icecaps.[7] An extreme example of global warming is the planet Venus, with its predominantly CO_2 atmosphere.[8] The temperature on Venus is 900 degrees Fahrenheit (482 degrees Celsius), high enough to melt lead.[9] Global warming may cause human health problems, extreme weather changes, sea level rise, adverse effects on agriculture, and degradation of the ecosystem.[10] Action is needed to prevent such devastating global

warming effects.

An increased average temperature will cause a higher number

of heat casualties.[11] A given rise in mean temperature above 60

degrees Fahrenheit is associated with more fatalities than a

similar decrease in temperature.[12] Global warming could also

result in an increase in diseases and deaths due to respiratory,

cardiovascular and cerebrovascular disease caused by pollution.[13]

In the 1998 heat wave, health care officials warned two-thirds

of the eastern United States of health care risks. In the heat

wave of 1995, the elderly, children, and asthmatics in

Washington, D.C. were encouraged to stay inside due to the

increase in air pollution. In Chicago alone, over 700 deaths

were attributed to the 1995 heat wave.[14]

The increased temperature would provide a more favorable

environment for the growth of bacteria. This growth would

inevitably be accompanied by bacterial mutation resulting in

antibiotic resistance. Also, a warmer climate would allow

disease-carrying insects to flourish.[15] For example, mosquitoes

carrying dengue fever have been found in the U.S. since 1986.

Cases of malaria have recently been found in airport personnel

in New York and London, transmitted by mosquitoes brought in by

commercial jets from tropical areas. There is no vaccine

against malaria and certain forms are resistant to almost all drugs.[16] A possible defense is genetic re-engineering aimed at disease-carrying tropical parasites and mosquitoes that are advancing due to global warming.[17]

Global warming can cause diverse disasters. Increases in temperature lead to abnormally hot, dry weather. The resultant devastating fires have governmental, commercial, industrial and human costs. Fires displaced more than 112,000 Florida residents from their homes in July 1998.[18] As the temperature has risen this past century, sea levels have also. Rising sea levels produce flooding, with population displacement and infrastructure damage.[19] A conservative current estimate is that flooding due to rising sea level will threaten 92 million people by the year 2100 if global warming is not halted.[20] Climate changes would affect growing seasons. With continued increase in temperature, some areas would receive less rainfall, while other areas would be saturated. Increased flooding would affect the quality and quantity of food production, especially in Third World countries. This would increase malnutrition and hunger, with subsequent impairment of childhood growth and development. Flooding-associated pollution could contaminate drinking water and spread life-threatening diseases, such as cholera.

Imbalances in the ecosystem also affect global warming. Rain forests filter and use the carbon dioxide from the air, which reduces global warming. Loss of the rain forests destroys the cleansing ability of the atmosphere, threatens wildlife, creates new semi-deserts and increases large-scale flooding.[21] Deforestation, especially by burning, contributes to an increase in CO_2 emissions.

Biological diversity is defined as the variety of plant and animal life that exists on the Earth or in some particular region of the Earth. The greater the diversity of organisms in an area, the healthier that region is likely to be.[22] Ninety percent of the planet's biological diversity is located in the tropics, with 60 percent in Latin America.[23] Rain forests make up only 7 percent of all woodlands on Earth, but contain millions of unique species. Tropical forests around the equator are only 7 percent of the Earth's dry land surface, but may contain 50 percent of all species of plants and animals.[24]

Tropical forests have been the source of 60 percent of the anticancer drugs discovered in the past 10 years. Depletion of the rain forests has caused the extinction of hundreds of species, a loss of potential biological resources for research into the development of new medicines. A current Brystol-Meyers Squibb Pharmaceutical Research Institute project seeks to demonstrate to developing nations the long-term economic benefit of preserving their rainforests.[25]

The Clinton administration has proposed a vigorous domestic program to research and develop methods for decreasing CO_2

emissions. It provides tax incentives to individuals and industry for using more energy efficient systems. A government/industry program called Partnership for Advancing Technology in Housing encourages more energy-efficient housing. If its 2010 goals are met, consumers will save $11 billion dollars annually. Also, carbon emissions will be reduced by 24 million tons annually, the amount currently produced by 20 million cars.[26]

In order to halt global warming, humans must reduce the combustion of fossil fuels, such as coal, oil, and natural gas. Transportation accounts for 80 percent of U.S. petroleum use. Using hydrogen fuel cells to replace turbine and internal combustion engines has great potential for reducing the harmful emissions. A fuel cell converts the chemical energy of fuel into usable electricity and heat without combustion as an intermediate step. Fuel cells provide inexpensive, clean power from hydrogen. They could become the principal energy source for electricity generation as well as for powering transportation. Although some fear that a shift to fuel cells would disrupt the oil industry and the world economy, fossil fuels could be utilized as the hydrogen source. The carbon residue that remains after extracting the hydrogen could be safely stored.[27]

Daimler-Chrysler Automobile Manufacturers recently announced the production of a car fueled by liquid hydrogen. Liquid hydrogen produces twice the energy of gasoline and has no harmful emissions. The production of the hydrogen fuel from fossil fuel is 70 to 90 percent cleaner than the current process from refining crude oil into gasoline. As the prototype burns fuel, the vehicle emits a small mist of water vapor and heat. Five thousand of these cars will be available by 2004, according to a spokesman. Ford and other U.S. automobile producers have similar plans.[28] California is working with a merged Daimler-Chrysler/Ford team to speed up the transition. The California Fuel Cell Partnership is a unique collaboration between Texaco, Shell, ARCO and the automobile industry. The partnership plans to place about 40 fuel cell passenger cars and 25 fuel cell buses on California roads between 2000 and 2003. Ballard Power Systems will provide the fuel cells for the vehicles and Texaco, Shell and ARCO will provide the fuel.[29]

THE KYOTO CONVENTION

Representatives of over 150 countries met in Kyoto, Japan December 1997 to address climate change. The Kyoto Protocol, produced by the conference and signed by the United States, is a commitment to lower greenhouse gases worldwide.[30] The U.S.

emits 22 per cent of the world's greenhouse gases. The former Soviet Union countries, Western Europe, China and Japan account for 14, 13, 10 and 5 percent, respectively.[31] Clinton will not ask Congress to ratify the Kyoto Protocol until China, India and other key developing nations agree to participate.[32]

Prior to the convention, the U.S. perspective was that in order to decrease emissions, people would have to adversely alter their lifestyles. After the Kyoto Convention, the U.S. government decided to emphasize energy efficiency to reduce energy consumption without lowering the standard of living. The Department of Energy has been directed to focus on research and development of technologies to lessen our dependence on current modes of transportation.[33]

Figure 1 shows that economic growth, energy consumption and carbon emissions are closely linked.[34]

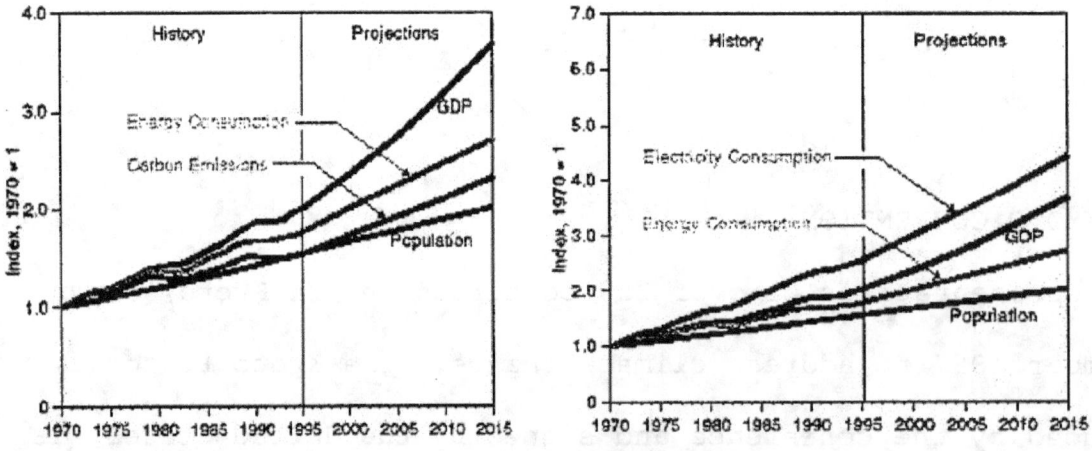

Figure 1. World Carbon Emission, Energy, GDP and
Population Trends

In developed countries, there is an approximate 0.75 percent increase in energy consumption for every 1 percent increase in GDP. In underdeveloped or developing countries there is a one to one ratio between GDP growth and increases in energy consumption. In the future, GDP will grow slightly faster than energy demand in developing countries. The challenge is how to promote economic growth without increasing carbon emissions. Currently, less polluting energy sources, such as solar panels, are too costly and would hinder economic growth.[35]

Energy consumption will continue to rise as populations grow and economies develop. Figure 2 shows that developing nations' energy consumption will surpass that of developed countries within the next 20 years.[36]

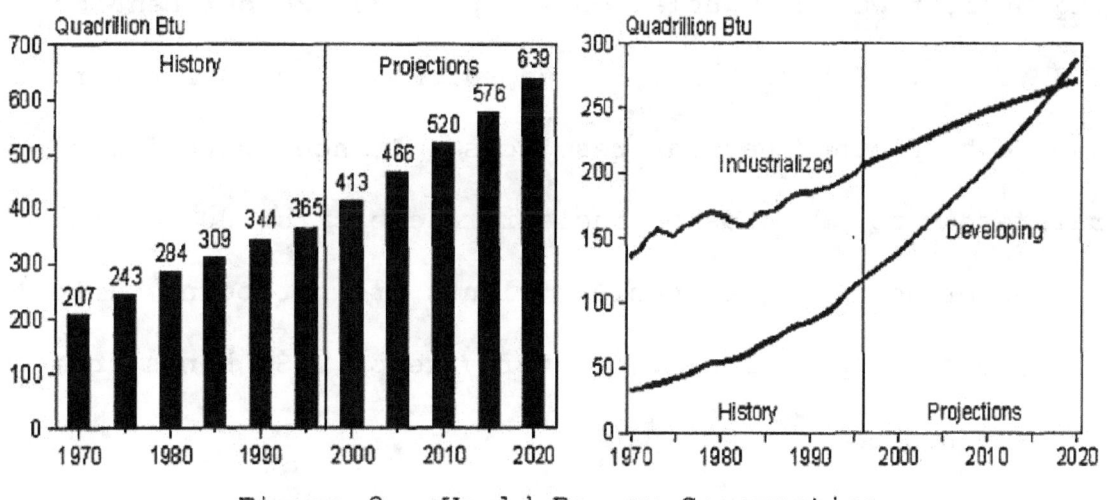

Figure 2. World Energy Consumption

One global approach is to share energy technology with the developing countries. For example, it has taken many stages or phases for the U.S. to research and develop better environmental technologies. The U.S. could offer these technologies to developing countries to speed their transition to environmentally suitable processes.[37]

Coal is abundant in China, India, and the United States. Although burning coal pollutes and produces greenhouse gases, China, India and the U.S. will not forgo this valuable resource unless there is a better replacement. One possibility is nuclear power. Nuclear reactors do not produce greenhouse emissions.[38] Safe nuclear power would be a viable replacement for burning coal. Currently, the debate on the use of nuclear power is more political and emotional than scientific. Scientists know of safe methods of producing nuclear energy and storing nuclear waste. Nuclear power produces 20 per cent of all U.S. energy. The U.S. is not dependent on nuclear power because we have other natural resources. France, with fewer natural resources, depends on nuclear reactors for 80% of its power. Unfortunately, new nuclear plants are not being built in the U.S., because political controversy keeps U.S. banks from funding any nuclear projects.[39]

Prior to 1995, most talks to address ozone depletion were limited to developed countries, since nearly 80 percent of all chloroflourocarbon (CFC) or ozone depleting gases are produced by the United Stated, the former USSR, Europe, Japan, and

China.[40] In December 1995, representatives from over 100

countries gathered in Vienna to review the protocol for managing

ozone depletion. They committed to restoring the ozone layer by

the year 2050. To achieve that, the three main chemicals that

deplete the ozone layer will be phased out. For example, the

target date for the phase-out of methyl bromide is 2010, with a

25 percent reduction by 2001 and 50 percent by 2005.[41]

ECONOMIC FACTORS

A conservative estimate is that a rise in mean temperature

of 2.5 degrees Celsius would cost the U.S. $60 billion annually,

1 percent of Gross Domestic Product (GDP). (See Table 1.) The

economic impact of mean temperature rise is not a linear effect.

on costs. For example, if the temperature rose 10 degrees

Celsius, the economic effect would increase six-fold to $360

billion dollars or 6 percent of GDP. The costs to other

Category	Costs in Billions*
Agricultural Losses	$18
Rise in Sea Level	$ 7
Electrical/ Air Conditioning	$11
Loss of Water Supply	$ 7
Increase in Urban Pollution	$ 4
Heat-Induced Deaths	$ 6
Deforestation	$ 3
Ski Industry Losses	$ 1.5
*1990 estimate	

TABLE 1. U.S. Economic Effects of Temperature Rise

countries, especially island countries at sea level, would be a greater percentage of their GDP.[42] The demand on resources to repair the damages created by global warming could impact negatively on the military budget.

Another economic factor that has affected defense spending is health care costs. Between 1940 and 1992, annual U.S. health care costs rose from $4 billion to $800 billion. From 1970 to 1998, health care costs rose from 6 percent to 13.4 percent of GDP, while defense spending dropped from 8 percent to approximately 3 percent of GDP.[43] If global warming continues, the health of children and future generations will be adversely affected due to the increasing pollutant load. Increasing health care cost at military facilities could impact readiness, with budget shifts required to pay for health care rather than training. Federal budget pressure from rising civilian health care costs could create more pressure for shifting funds from military to health programs.

THE ENVIRONMENT AND THE AMERICAS

Since the end of the Cold War, U.S. interests have increasingly focused on Canada and Latin America. In Latin America and the Caribbean, 34 of the 35 countries are practicing democracy, with Cuba the exception. For these democracies to flourish, they need economic growth and low unemployment. U.S. importation of raw materials and goods produced in countries with lax environmental standards has contributed to Third World environmental degradation. Now the U.S. wants to bolster emerging democracies, and encourage international environmental strategies that will promote Latin American security, stability, and sustainability.[44] Latin American countries are keenly aware that, while pushing other countries for environmental reform, the United States produces most of the ozone depleting gases. Latin America produces less than 3 percent of the degrading gases, with two-thirds of that from Brazil and Mexico.[45]

Canada is concerned about the affects of global warming:

> "Global warming causes high temperature, droughts, and rises in tide resulting in increased cost of food, water and land. Animals will be either endangered or extinct due to heat, which can destroy or alter habitat. Cities on or close to coasts will be destroyed because of the rise in tide due to melting of icebergs in the North."[46]

On the other hand, the quality of life could be adversely affected by decreased industrialization and decreased use of cars. Canada has been economically affected by two recent major floods and is seeking ways to prevent this trend.[47] Another issue in Canada is deforestation. The government regulates the harvesting of trees in Canada. For every tree removed the buyer must plant three that thrive for five to ten years.[48] Canada has temporarily halted the export of drinking water. Previously,

13

water was exported without environmental assessment. Canada will resume water exportation once the government sets specific guidelines.[49] The debate in Canada, as in most countries, is how much of the budget should be allotted to improve environmental security.

MILITARY IMPLICATIONS OF GLOBAL WARMING

Originally, the U.S. military was concerned the Kyoto Treaty would limit training and thus reduce readiness. That need not be a concern.

> "Both multilateral and unilateral operations are protected under the Kyoto Protocol. When our military forces conduct or train for operations overseas, the greenhouse gases do not count against U.S. limits for several reasons. First, under the Kyoto agreement, emissions count against the country in which they occur (in unilateral and multilateral training and operations). Second, the Kyoto agreement exempts military emissions in international airspace and on high seas (in unilateral and multilateral training and operations). Third, the Kyoto agreement exempts emissions from multilateral military operations."[50]

The Federal Government is the largest energy user in the United States. Since the Government spends $8 billion annually on energy costs, energy efficient technologies are being placed in federal facilities to reduce cost and emissions. Executive Order 12902 requires DOD to reduce energy usage 30 percent between 1985 and 2005. DOD reduced energy expenditures by $450 million in 1997 and consumed 17 percent less energy than in

1985.   Under the Executive Order, DOD must reduce energy use by another 13 percent by 2005.[51]

Methods DOD is employing to reach this goal are numerous. The U.S. Army Corps of Engineers has a newly established program, the Defense Energy Support Center, which assists installations in all 50 states, the District of Columbia and Puerto Rico with Energy Savings Performance Contracts (ESPCs). The ESPCs will require minimal funds to decrease energy expenditures.   DOD is using equipment and materials that are in the top 25 percent of energy efficiency.   Solar sources are being installed to heat water in 652 family housing units in Fiscal Year 1999.   Future building projects will use energy efficient designs, proven to reduce energy use 30 to 50 percent with minimal implementation costs.   These efforts by DOD will lower costs, enhance military readiness, and reduce emission of greenhouse gases.[52]

Increased flooding and natural disasters have a destabilizing effect on foreign countries.   The size of the area destroyed directly affects the response of the host nation.   If agricultural areas and/or jobs are lost, there tends to be migration to another area or country for food and jobs.   This can result in instability, which may require U.S. military intervention to resolve.   Changing climate can lead to increased flooding and fires, which will increase the need for humanitarian assistance, often involving the U.S. military.

The effect of global warming on the health of soldiers and future recruits is a concern. Asthma is a condition that disqualifies individuals for military service.[53] Global warming has been accompanied by a rise in childhood asthma, because the higher temperatures increase pollution and smog.[54] Continued rising temperatures will only worsen existing conditions and create new cases of respiratory illnesses.

The current generation of children has been exposed to a more toxic environment than children in the past. This environmental exposure is thought to be responsible for a rise in chronic illnesses.[55] The Environmental Protection Agency has published a report on plans to protect children from environmental threats:

> "Throughout the world, children face significant threats from an array of environmental hazards. They may absorb some pollutants more rapidly and eat more foods, drink more liquids and breathe more air in proportion to body weight than the typical adult. Their neurological, immunological, reproductive, digestive, and other bodily systems are still developing, providing windows of vulnerability for adverse effects. Children are less able to recognize and protect themselves from exposure to environmental pollutants, and childhood activities put them in closer contact with environmental hazards. By virtue of their youth, children exposed to environmental pollutants have a long period during which latent effects may become manifest."[56]

The Army is required to comply with environmental statutes, to include:

- Installation, operation, and maintenance of air and water pollution control technology.
- Quantitative and qualitative limitations on air and water emissions.
- Pollution monitoring, record keeping, and reporting requirements.
- Operating permits for pollution sources and the payment of reasonable permit fees.
- Handling, transportation, storage, treatment, and disposal of solid waste and hazardous waste.
- Reporting and cleanup of spills.
- Monitoring underground storage tanks for leaks.
- Noise control.
- Cleanup of active and closed hazardous waste disposal sites
- Conservation of endangered and threatened species and wetlands.[57]

In 1990, legislation was passed stating that environmental protection is the "primary mission in the planning, design, construction, operating and maintenance of water sources projects--along with navigation and flood control."[58] This legislation expanded the role of the Corps of Engineers. The Corps is responsible "for restoration of fish and wildlife habitat, by including mitigation in the design of all its projects, by protecting environmental assistance, such as wetlands through its regulatory program, and by its program of environmental compliance at Civil Works project sites."[59]

17

CONCLUSIONS

Environmental issues are an increasing U.S. priority because
of recognition that the quality of the environment affects the
quality of life. The 1998 National Security Strategy devoted
four times more space to environmental security than did the
previous year.[60] The domestic and global environmental goals of
the Clinton Administration are far-reaching. Meeting these
goals will positively affect American and global societies.
Economic downturns could weaken funding for the environment.
However, investments to reduce global warming will bring long-
term benefits to human lives and the economy. Clinton will
propose a $4 billion environmental budget for Fiscal Year 2000--
$200 million dedicated to a "clean air partnership" to reduce
greenhouse gases caused mainly by the production of CO_2 from
burning fossil fuels.[61] Much of the proposed budget, over $1.4
billion, will provide tax incentives for developing energy
efficient technology. Vice President Gore says these funds
"help protect public health."[62] The administration must justify
the requested funds necessary to achieve its goals.

As a senator, Gore wrote in 1992 of the complexity of
solving global environmental concerns. He suggested the need
for a "Global Marshall Plan," referring to the 1947 European
Recovery Program that Marshall and Truman wrote to help rebuild

Western Europe and thwart communist expansion. Gore suggested

six strategic goals to accomplish this:

- stabilizing the world's population

- rapidly creating and developing environmentally
  appropriate technologies

- comprehensively changing the economic rules by which
  we measure the impact of our decisions on the
  environment

- negotiating and approving a new generation of
  international agreements to decrease global warming

- educating the world's citizens about the global
  environment

- establishing social and political conditions in the
  developing world conducive to the emergence of
  sustainable societies[63]

Some of the goals are more achievable than others.

World population is predicted to continue to grow.

Fortunately, technologies are being developed to utilize

cleaner fuel systems. Agreements to share energy

technology and other efforts to reduce global warming could

be made at international conferences. The Kyoto Protocol

is an initial effort; China and India may sign if they can

realize economic and environmental benefits. There is a

need to educate the world on environmental concerns. As

the following cartoon shows, not all Americans are

enlightened on the subject.

In 1992, Gore blamed consumers for destroying the Earth and wanted the developed countries to implement radical lifestyle changes. Since the 1997 Kyoto Conference, the emerging theme has switched from changing lifestyle to leveraging technology. This appears to be more palatable to society.[65]

In Environmental Gore, a group of experts in various fields evaluated, praised and criticized Vice President

Gore's views on environmental improvements. Economists and scientists agree greenhouse gases are degrading the environment, but recommend less costly methods to decrease emissions. They recommend academic and analytical solutions, not just political ones.[66] Perhaps these experts could be assembled to develop Gore's environmental "Marshall Plan." Both political parties should be represented for mutual agreement on the ends, ways, and means to solve the problem.

The Army Corps of Engineers could work jointly with developing nations to make waterways less susceptible to flooding; and to rebuild roads, schools, and health clinics. Initially, the focus of the military should be in Brazil and Mexico, the leading Latin American contributors to global warming. Goals could include decreasing CFC gases, reforestation, and teaching hazardous waste management.

Finally, the U.S. military should become a leader in compliance with federal, state, and local environmental laws. DOD and experts in the field of environmental security should work together to ensure that specific and mutual goals support the best interests of the United States. More specific goals (the ends) need to be outlined by the Department of Defense to give better guidance (the ways) for what the Army needs to accomplish in the future. The DOD should provide a budget (the means) for the Army to accomplish the environmental mission. By supporting environmental issues, the military could be used in preventive roles that are less costly than reconstruction and reconstitution.

With a decrease in global warming there would be less pollution-induced disease. Military medicine could focus more on preventing rather than treating diseases. Engineer units could assist in constructing adequate roads, bridges, schools, and health clinics, rather than reconstruction during crisis management. The long-term goal is for the U.S. to be a leader in global environmental stability, so the U.S. can concentrate more efforts on domestic issues.

Most do not view global warming as a military issue, but global warming and other environmental issues have important implications for the U.S. military. As the United States leadership in environmental matters encourages global stability, the U.S. military will be able to focus more on readiness, training, and operations.

Word Count: 4239

ENDNOTES

[1] White House, <u>A National Security Strategy for a New Century,</u> (Washington: U.S. Government Printing Office, October 1998), 13-14.

[2] "White House Climate Control Fact Sheet," June 1998; available from http://www,usai.gov/topical/global/environ/whfct698.htm; Internet; accessed 8 October 1998.

[3] Ibid.

[4] Rafe Pomerance, "Testimony on Ozone Depletion USIS," 25 January 1996; available from <http://usai.gov/topical/global/environ/promerance.htm>; Internet; accessed 8 October 1998.

[5] Astrid Zwick. "Global Climate Change: Potential impact on Human Health;" available from <http://www.jrc.es/iiiiptsreport/vol13/english/lif11E136.htm>; Internet; accessed 24 February 1999.

[6] "White House Climate Control Fact Sheet," June 1998.

[7] Richard Hillmand, <u>Understanding Contemporary Latin America</u> (Boulder, London: Lynne Rienner Publishers, 1997), 198.

[8] Jim Fuller, "Global Warming: The Consequences for life on Earth;" April 1997; available from <http://www.usai.gov.journals/itge/gi-3.htm>; Internet; accessed 8 October 1998.

[9] "Venus Facts;" available from <http://202.102.153/collect/sun/Venus/HTML/index.html>; Accessed 5 April 1999.

[10] Fuller.

[11] Timothy Wirth. "Why We Should Care," April 1997; available from <http://www.usai.gov/journals/itgiic/047/ijge/gj-4.htm>; Internet; accessed 8 October 1998.

[12] "EPA Global Warming: Impacts—Health," 24 January 1999; available from <http://www.epa.gov/oppeoee1/globalwarming/impacts/health/index.html>; Internet; accessed 19 April 1999.

[13] "Social and Economic Effects of Global Warming," available from <http://www.chem.cm.edu/courses/chem105/projects/group1/page1/html.>; Internet; accessed 24 February 1999.

[14] "Heat Wave Grips Much of United States,"26 June 1998; available from <http://www.crosswalk.com/reuters/nefile5.htm>; Internet; accessed 17 March 1999.

[15] Judith Randal, "A Paradise for Pathogens—Almost Everywhere," November 1996; available from <http://www.usai.gov/journals/itgic/1196/itje/gi-4.htm>; Internet; accessed 8 October 1998.

[16] Ibid.

[17] Mark Moran, "Genetic Engineering Advances Now Targeting Malaria Bugs," Health and Science 42 (February 1999): 35-37.

[18] Patrick Mazza, "Going to Blazes," 7 July 1998; available from <wysiwyg://11/http://bsd.motherjones.com/news_wire/mazza.html>; Internet; accessed 15 March 1999.

[19] Zwick.

[20] Wirth.

[21] John L. Peterson, The Road to 2015: Profiles of the Future. (Corte Madera,CA: Waite Group Press: 1994), 82.

[22] David Newton, Global Warming (Santa Barbara, CA: ABC-CLIO, 1993), 147.

[23] Hillmand, 200.

[24] Thomas J. Lovejoy, "Biodiversity: The Most Fundamental Issue," 1 March 1994; available from <http://www.erin.gov.au/portfolio/esd/biodiv/articles/lovejoy.html>; Internet; accessed 7 October 1998.

[25] "Can Economics Save Suriname Rainforest?" 7 April 1998; available from <http://www.ennccom/news/enn-stories?1998/04/040798/suriname.asp>; Internet; accessed 10 May 1999.

[26] "White House Climate Change Fact Sheet," June 1998.

[27] R. James Caverly, Office of Science and Technology Policy, U.S. Department of Energy, PowerPoint® briefing slides provided during interview by author 26 March 1999, Carlisle, PA.

[28] Cable News Network Broadcast, 12 March 1999.

[29] Cable News Network Broadcast, 23 April 1999.

[30] Michael Tebo, "U.S. Signs Kyoto Protocol," 12 November 1998; available from <http://www.weathervane.rff.oorg/negtable/US_signs.html>; Internet; accessed 5 April 1999.

[31] Hillmand, 200.

[32] Tebo.

[33] Caverly.

[34] Ibid.

[35] Ibid.

[36] Ibid.

[37] Ibid.

[38] John A. Baden, Environmental Gore (San Francisco: Pacific Research Institute for Public Policy, 1994), 197-198.

[39] Kent H. Butts, United States Army War College George C. Marshall Professor of Military Studies, interview by author, 29 March 1999, Carlisle, PA.

[40] Hillmand, 200.

[41] Ibid.

[42] William R. Cline, Global Warming: The Economic Stakes (Washington: Institute for International Economics, 1992), 49.

[43] Peterson, 220-221.

[44] L. Erik Kjonnerod, Evolving U.S. Strategy for Latin America and the Caribbean: Mutual Hemisphere Concern and opportunities for the 1990's (Washington: National Defense University Press, 1992), 116-117.

[45] Hillmand, 200.

[46] "Canada: What We've Learned"; available from <http://www.hisurf.con/~lena/global/learned.html>; Internet; accessed 9 March 1999.

[47] Ibid.

[48] COL Glenn Nordick, Canadian Forces, interview by author, 9 March 1999, Carlisle, PA.

[49] Ibid.

[50] Sherry W. Goodman, "Kyoto Treaty Doesn't Compromise Our National Security," 6 June 1998; Available from <http://www.denix.osd.mil/denix/Public/News/OSD/Climate/washtimes.html>; Internet; accessed 24 February 1999.

[51] Ibid.

[52] Ibid.

[53] Department of the Army, Standards of Medical Fitness, Army Regulation 40-501 (Washington, D.C.: U.S. Department of the Army, 30 August 1995), 9.

[54] Zwick.

[55] "Earth Link: March/April 1997;" available from <http://www.epa.gov/earlink1/earthlink/97marapr.htm>; Internet: accessed 28 October 1998.

[56] Ibid.

[57] How the Army Runs: A Senior Leader Reference Handbook, (Carlisle Barracks, PA: U.S. Army War College, 1997) 21-7, 21-8.

[58] Ibid.

[59] Ibid.

[60] National Security Strategy, 13-14.

[61] "Clinton Wants $4 Billion to Fight Global Warming," Carlisle (PA) Sentinel 26 January 1999, sec. A, p.3.

[62] Ibid.

[63] Senator Al Gore, Earth in the Balance: Ecology and the Human Spirit (Boston, New York, London: Houghton Mifflin Company, 1992) 305-307.

[64] MacNelly, International Herald Tribune (political cartoon), 19 December 1997.

[65] Caverly.

[66] Baden, xv-xvi.

BIBLIOGRAPHY

Baden, John A. Environmental Gore: A Constructive Response to
    Earth in the Balance. San Francisco: Pacific Research
    Institute for Public Policy, 1994.

Butts, Kent H., George C. Marshall Professor of Military
    Studies, United States Army War College. Interview by
    author, 29 March 1999, Carlisle, PA.

Cable News Network, 19 March 1999.

Cable News Network, 23 March 1999.

"Canada: What We've Learned." Available from
    <http://www.hisurf.conn/~lena/global/learned.html>.
    Internet. Accessed 9 March 1999.

"Can Economics Save Suriname Rainforest?" 7 April 1998.
    Available from <http://www.ennccom/news/enn-
    stories?1998/04/040798/suriname.asp>. Internet. Accessed
    10 May 1999.

Caverly, R. James, Office of Science and Technology Policy,
    U.S. Department of Energy, PowerPoint® slides provided during
    interview by author 26 March 1999, Carlisle, PA. Cline,
William R. Global Warming: The Economic Stakes. Washington,
D.C.: Institute for International Economics, 1992.

"Clinton Wants $4 billion to fight Global Warming," Carlisle
    (PA) Sentinel, 26 January 1999, sec. A, p.3.

 "Earth Link: March/April 1997." Available from
    <http://wwww.epa.gov/earlink1/earthlink/97marapr.htm>.
    Internet. Accessed 28 October 1998.

"EPA Global Warming: Impacts--Health." Updated 24 January, 1999.
    Available from <http://www.epa.gov/oppeoee1/global
    warming/impacts/health/index.html>. Internet. Accessed
    19 April 1999.

Fuller, Jim "Global Warming: The Consequences for Life on
    Earth." April 1997. Available from
    <http://ww.usai.gov.journals/itge/gi-3.hhtm>. Internet.
    Accessed 8 October 1998.

Goodman, Sherry W.   "Kyoto Treaty Doesn't Compromise Our
     National Security." 6 June 1998. Available from
     <http://www.denix.osd.mil/denix/Public/News?OSD/Climate/wash
     times.html>. Internet. Accessed 24 February 1999.

Gore, Al. Earth in the Balance: Ecology and the Human Spirit.
     Boston, New York, London: Houghton Mifflin Company, 1992.

How the Army Runs: A Senior Leader Reference Handbook.
     Carlisle, PA: U.S. Army War College, 1997.

"Heat Wave Grips Much of United States." 26 June 1998.
     Available from
     <http://www.crosswalk.com/reuters/nefile5.htm>. Internet.
     Accessed 17 March 1999.

Hillmand, Richard S.   Understanding Contemporary Latin America.
     Boulder, London: Lynne Reinner Publishers, 1997.

Kjonnerof, L. Erik.   Evolving U.S. Strategy for Latin America
     and the Caribbean: Mutual Hemispheric Concerns and
     Opportunitis for the 1990's. Washington, National Defense
     University Press, Fort Lesley J. McNair, 1992.

Lovejoy, Thomas J. "Biodiversity: The Most Fundamental Issue."
     1 March 1994. Available from
     <http://www.erin.gov.au/portfolio/esd/biodiv/articles/lovejo
     y.html>. Internet. Accessed 7 October 1998.

MacNelly. "International Herald Tribune." Political Cartoon.
     19 December 1997.

Mazza,Patrick. "Going to Blazes." 7 July 1998. Available from
     <wysiwyg://11/http://bsd.motherjones.com/news_wire/mazza.htm
     l>. Internet. Accessed 15 March 1999.

Moran, Mark. "Genetic Engineering Advances Now Targeting
     Malaria Bugs." Health and Science. Vol.42, no.822.
     February 1999: 35-37.

Newton, David E. Global Warming: A Reference Handbook. Santa
     Barbara: Instructional Horizon, Inc., 1993.

Nordick, Glen, Colonel, Canadian Forces. Interview by author,
     9 March 1999, Carlisle, PA.

Petersen, John L. The Road to 2015: Profiles of the Future. Corte Madera: Waite Group Press, 1994.

Pomerance, Rafe, "Testimony on Ozone Depletion USIS." 25 January 1996. Available from <http://www.usai.gov/topical/global/environnn/pomerance.htm> Internet. Accessed 8 October 1998.

Randal, Judith. "A Paradise for Pathogens—Almost Everywhere," November 1996. Available from <http://www.usai.gov.journals/itgic/1196/itje/gi-4.htm.> Internet. Accessed 8 October 1998.

"Social and Economic Effects of Global Warming." Available from <http://www.chem.cm.edu/courses/chem105/projects/group1/page 1/html.>. Internet. Accessed 24 February 1999.

Tebo, Michael. "U.S. Signs Kyoto Protocol." 12 November 1998. Available from <http://www.weathervane.rff.oorg/negtable/US_signs.html>. Internet. Accessed 5 April 1999.

U.S. Department of the Army. Standards of Medical Fitness. Army Regulation 40-501. Washington, D.C.: U.S. Department of the Army, 30 August 1995.

"Venus Facts." Available from <http://202.102.153/collect/sun/Veuns/HTML/index.html>. Accessed 5 April 1999.

White House, A National Security Strategy for a New Century, (Washington: U.S. Government Printing Office, October 1998), 13-14.

"White House Climate Control Fact Sheet." June 1998. Available from <http://www,usai.gov/topical/global/environ/whfct698.htm>. Internet. Accessed 8 October 1998.

Wirth, Timothy. "Why We Should Care." April 1997. Available from <http://www.kusaiiii.gov.journals/itgiic/047/ijge/gj-4.htm>. Internet. Accessed 8 October 1998.

Zwick. "Global Climate Change: Potential Impact on Human Health." <Available from <http://www.jrc.es/iptreport/vol13/english/lif11E136.htm>. Internet. Accessed 24 February 1999.